WONDERS OF THE WORLD: MOTHER NATURE AT WORK

Speedy Publishing LLC
40 E. Main St. #1156
Newark, DE 19711
www.speedypublishing.com

Copyright 2018

All Rights reserved. No part of this book may be reproduced or used in any way or form or by any means whether electronic or mechanical, this means that you cannot record or photocopy any material ideas or tips that are provided in this book

OUR EARTH IS FULL OF WONDERS
CARVED OUT BY NATURE.

Iguazu Falls are waterfalls located on the border of Argentina and Brazil. The falls are 1.7 miles long and with more than 275 falls, the Iguazu are the most majestic of water falls.

Jeju Island is the largest island off the coast of the Korean Peninsula. It is a volcanic island dominated by Hallasan, a volcano 1,950 metres high and the highest mountain in South Korea.

Komodo island is one of the 17,508 islands that compose the Republic of Indonesia. The island is popular for tourists who wants to see the komodo dragon, as well as divers who are eager to see the coral reefs in the area.

The Puerto Princesa Underground River is a protected area located in the Philippines. The river is 8.2km long and flows underground through natural rock formations such as caves, stalactites, and stalagmites.

Table Mountain is a flat-topped mountain overlooking the city of Cape Town in South Africa. The rocks of the mountain are approximately 600 million years old. The flat top of the mountain is often covered by orographic clouds.

Halong Bay is located in Quang Ninh province of Vietnam. The bay is famous for its 1,960–2,000 islets, most of which are limestone. This limestone has been forming for over 500 million years.

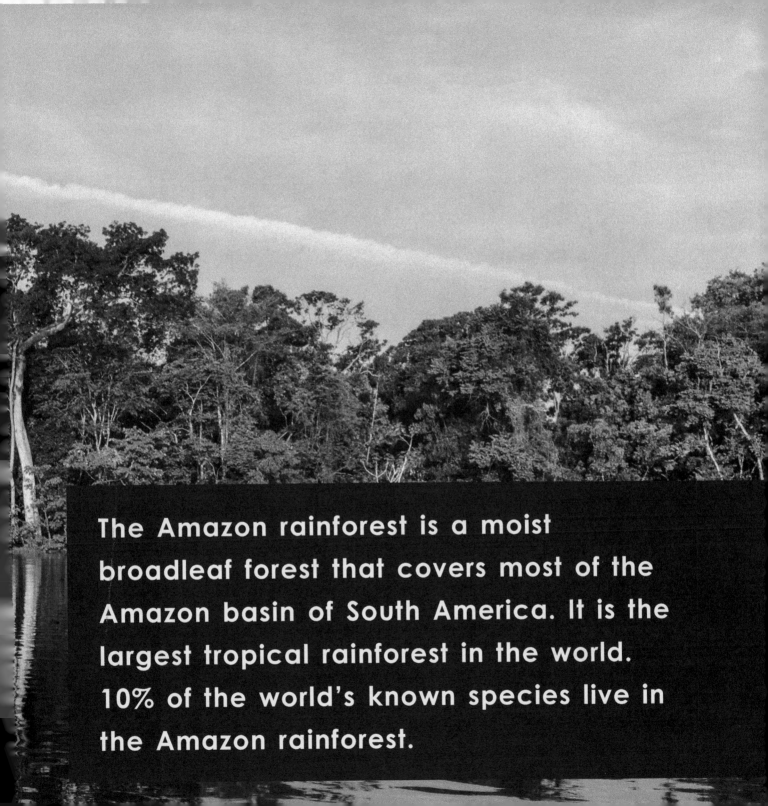

The Amazon rainforest is a moist broadleaf forest that covers most of the Amazon basin of South America. It is the largest tropical rainforest in the world. 10% of the world's known species live in the Amazon rainforest.

Printed in the USA
CPSIA information can be obtained
at www.ICGtesting.com
LVHW081109161123
763986LV00085B/2585

9 781682 801178